How does a PLANT become OIL?

Linda Tagliaferro

www.raintreepublishers.co.uk
Visit our website to find out more information about Raintree books.

To order:
☎ Phone +44 (0) 1865 888066
▤ Fax +44 (0) 1865 314091
▣ Visit www.raintreepublishers.co.uk

Raintree is an imprint of Capstone Global Library Limited, a company incorporated in England and Wales having its registered office at 7 Pilgrim Street, London, EC4V 6LB – Registered company number: 6695582

"Raintree" is a registered trademark of Pearson Education Limited, under licence to Capstone Global Library Limited

Text © Capstone Global Library Limited 2010
First published in hardback in 2010
The moral rights of the proprietor have been asserted.

Edited by David Andrews and Laura Knowles
Designed by Richard Parker and Wagtail
Original illustrations © Capstone Global Library Ltd 2010
Illustrated by Jeff Edwards
Picture research by Hannah Taylor and Sally Claxton
Originated by Modern Age Repro House Ltd
Printed and Bound in China by CTPS

ISBN 978 1 406211 23 8 (hardback)
14 13 12 11 10
10 9 8 7 6 5 4 3 2 1

British Library Cataloguing in Publication Data

Tagliaferro, Linda
How does a plant become oil?. – (How does it happen?)
572.2
A full catalogue record for this book is available from the British Library.

Acknowledgements
We would like to thank the following for permission to reproduce photographs: Akg-images p. **12** (North Wind Picture Archives); Alamy pp. **9** (© Derek Croucher), **16** (© Ace Stock Ltd), **17** (© Peter Jordan), **25** (© Trip), **28** (© Andrew Butterton); Corbis pp. **4** (Ed Kashi), **7** (zefa/ Dietrich Rose), **10** (Comstock), **13** (Underwood & Underwood), **15** (George Steinmetz), **20** (zefa/ Ted Levine), **22** (Latin Stock), **23** (zefa/ Alexander Benz), **26** (Jim Zuckerman), **27** (Martin Harvey); istockphoto **background image** (© Dean Turner); Science Photo Library pp. **5** (Lepus), **18** (Paul Rapson), **19** (BSIP, FIFE), **29** (SIMON FRASER).

Cover photographs of a tropical fern (top) reproduced with permission of istockphoto/©David Gunn and buckets of oil sludge from an oil spill (bottom) reproduced with permission of Corbis/epa/Alberto Estevez.

Every effort has been made to contact copyright holders of any material reproduced in this book. Any omissions will be rectified in subsequent printings if notice is given to the publisher.

Disclaimer
All the Internet addresses (URLs) given in this book were valid at the time of going to press. However, due to the dynamic nature of the Internet, some addresses may have changed, or sites may have changed or ceased to exist since publication. While the author and Publishers regret any inconvenience this may cause readers, no responsibility for any such changes can be accepted by either the author or the Publishers.

Contents

Some words are shown in bold, **like this**. You can find out what they mean by looking in the glossary.

Oil from plants

We use many products made from oil in our daily lives. In many ways, we depend on oil for our comfort and our transport. Oil products heat homes and provide fuel for cars, boats, and aeroplanes. They are also used to make all kinds of other things, from plastics to clothing. But where does oil come from?

This oil platform sits off the coast of Nigeria, Africa, in the Atlantic Ocean.

The story of oil goes back millions of years ago to **microscopic** (tiny) plants called **plankton**. These plants float on top of oceans and other bodies of water. There are thousands of types of these tiny plants.

Plankton drift with the movements of the water in which they live. After a long time, these plants become a type of oil we call **petroleum** or **crude oil**.

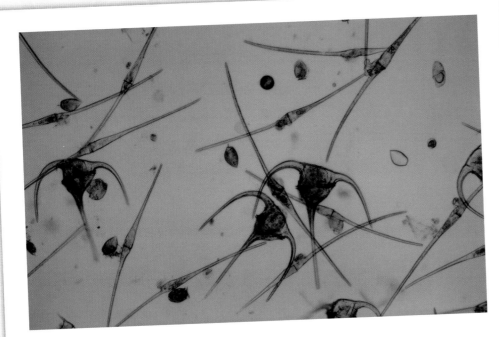

Magnified image of plankton

Tiny plankton float on top of the world's oceans.

Little drifters

The word plankton comes from a Greek word meaning "wanderer" or "drifter", because plankton drift with the ocean waves.

Tiny food "factories"

Tiny plants like **plankton** have something in common with large plants. They contain a special green substance called chlorophyll. This makes it possible for plants to make their own food.

To produce food, plankton first take energy from sunlight. The Sun only shines on the top of the oceans. Plankton must therefore live on the ocean's surface to catch this light. To make food, plankton also need a gas called carbon dioxide from the air and nutrients from the ocean.

Types of plankton

Most plankton are plants called phytoplankton (say "FIE-toe-PLANK-tun"). However, some kinds of plankton are actually animals. These are called zooplankton (say "ZOH-oh-PLANK-tun").

Plankton are very important for everything on Earth. This is because when these plants are finished making food, they give off the gas oxygen. People and animals need to breathe oxygen to stay alive.

Plankton get energy from sunlight shining on the ocean's surface.

From mud to rocks

Today's oil comes from **plankton** that died millions of years ago. When they died, **bacteria** (**microscopic** living things) broke down their tiny bodies. Their remains slowly sank to the bottom of the ocean. After many years, layers of sand and mud built up on the ocean floor, covering these remains.

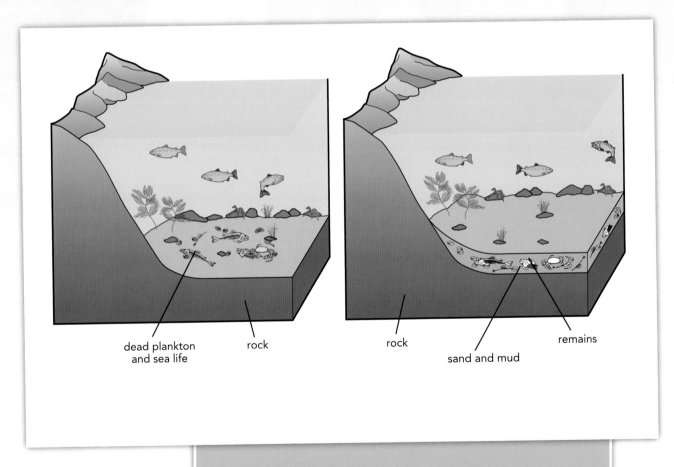

dead plankton and sea life

rock

rock

sand and mud

remains

Dead plankton and other sea life sink to the bottom of the ocean. Sand and mud cover the remains.

Starting the food chain

Plankton still live in our oceans and other bodies of water today. One reason why they are so important is because they are at the bottom of the **food chain**. In a food chain, plants are eaten by animals, and those animals are eaten by other, often larger, animals. Some animals and fish feed on plankton. Some larger animals feed on smaller fish that eat plankton, and so on. Every living thing in the ocean depends on the food chain that starts with plankton.

After millions of years, the sand and mud eventually turned into a type of rock, called **sedimentary rock**, that surrounded the plankton. The sedimentary rocks on top of the plankton were solid and very heavy. The remains of the plankton became trapped inside these rocks.

These cliffs are made of limestone, a sedimentary rock.

9

Energy from the Sun

Almost all energy, including energy released by burning oil, begins with the Sun. This is because plants such as **plankton** use the Sun's heat and rays to help them make their own food. That energy moves up the **food chain** when animals eat the plants.

Plants need the Sun's energy to help make their own food.

When plankton die without being eaten, they still have some energy from the Sun. That energy can be stored in Earth for millions of years. It is the energy found in oil.

How does this happen? When the plankton remains are buried inside rocks, the rocks' weight creates pressure and heat. After millions of years, the pressure and heat turn the plankton into a type of oil. We call this **crude oil** or **petroleum**. The word petroleum comes from two Latin words meaning "rock" and "oil". Today, we can use the energy from petroleum that was created in ancient times.

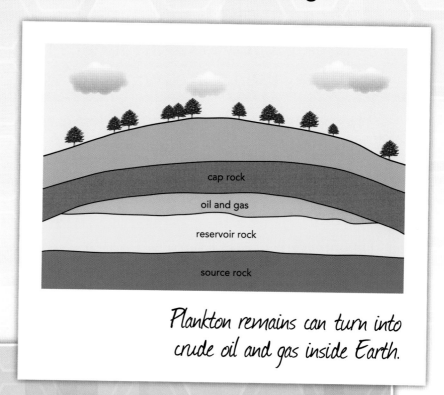

Plankton remains can turn into crude oil and gas inside Earth.

From plankton to natural gas

Natural gas also formed when plankton died and sank to the ocean floor. Natural gas has many uses, like heating houses and powering clothes dryers, water heaters, and ovens. **Bacteria** broke down the dead plankton, causing them to decay. During this process, natural gas was created.

Past uses of oil

Some oil can travel from deep inside Earth to the surface through cracks in rocks. However, most of it is trapped within rocks underground. Ancient civilizations discovered this sticky, waterproof liquid when it seeped to Earth's surface thousands of years ago. They used it in a number of ways.

The ancient Chinese drilled their own **oil wells**. An oil well is a deep hole in the ground from which oil can be drawn out. They used bamboo pipes to transport the oil. Ancient Egyptians used **asphalt** (a very thick oil) for preserving mummies. While fighting the ancient Greeks, the ancient Persians dipped their arrows in oil. They lit them and then shot the flaming arrows at their enemies.

During the 1100s, the Byzantine Empire used machines like this one to shoot burning weapons at their enemies.

The first oil well in Titusville, USA, 1861

This photograph shows Edwin Drake (on the right) and a friend standing near the oil well that Drake designed in what would become "Oil City".

Oil city

In 1859 a man called Edwin Drake, working for an oil company, set up an oil well near Titusville, Pennsylvania, USA. On 27 August 1859, the well struck oil, starting the oil industry in the United States. Others saw an opportunity. Within a little more than a year, 75 oil wells were set up. The area around these oil wells became known as "Oil City".

Digging deep

Oil was formed under oceans or lakes many millions of years ago. But Earth today is not the same as it was then. Over time, the continents moved. Areas that were underwater became dry land. This means that some oil that formed under ancient oceans could now be under land.

How do oil companies know where to drill to find these pockets of oil? First, their scientists and engineers study rocks on land or on the bottom of the ocean. Then they send special sound waves through the ground.

Sound travels at different speeds as it moves through different types of rock. Computers use this information to make a map of what is below the surface.

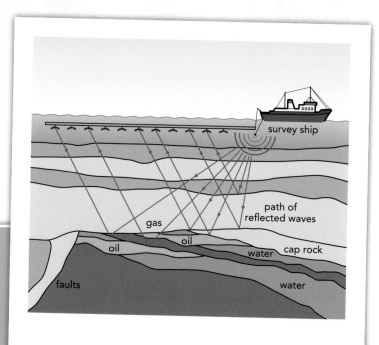

survey ship

path of reflected waves

gas

oil

oil

water

cap rock

faults

water

Sound waves help oil companies to work out where to look for oil.

An oil exploration rig in the North Sea

Exploration wells can be on land or, like this one, in the ocean.

Exploring for oil

Before setting up an **oil well**, oil companies drill exploration wells. They build these after scientists tell them where there might be oil. Exploration wells allow them to take samples of rocks from under the ground. This helps them to find out more about what might lie under Earth's surface.

Oil!

An oil derrick

When scientists find land that might contain oil, a company sets up an **oil well**. It includes a tall tower called an **oil derrick**. Inside the oil derrick is a very long pipe called a drill string that rapidly turns a drill bit. The drill bit is a strong cutting tool. It digs holes deep into the ground. Different drill bits are used for grinding holes in different types of rock.

A tall tower called an oil derrick holds a long pipe that turns a drilling tool.

A big metal pipe called a casing is put inside the long drilled hole. The casing keeps the hole open. Finally, a pump sucks the oil up through a tube and brings it to the surface.

This worker is checking a big drill bit in an oil rig.

Under the ocean

When oil companies drill for oil trapped under the ocean floor, they build special platforms to hold a **drilling rig**. This is a machine that drills holes into the ground. For drilling in very deep water, they use special rigs on ships. These rigs can drill an oil well in water up to 3,000 metres (10,000 feet) deep!

What is crude oil?

When oil first comes out of the ground, it is called **crude oil** or **petroleum**. It will later be heated and **processed** (changed) so that it can be used as fuel for cars, boats, and aeroplanes. It will also be made into other products, such as wax and nylon.

This crude oil has just come out of the ground. It will later be changed into different products.

Crude oil

Crude oil is a sticky liquid. It can be dark or almost colourless, depending on where it comes from. It is flammable, which means it can easily be set on fire. When crude oil first comes out of a well, it might contain water, sand, or gases. The gases are removed before the crude oil is sent to a place called a **refinery**. At the refinery, crude oil is processed into useful products.

Crude oil can be transported in an oil tanker like this one.

Measuring oil

We measure liquids like water and juice in litres and millilitres. But we use the term barrel as a measurement for crude oil. One barrel equals 159 litres (35 gallons).

At a **refinery**, **crude oil** is broken into different products. First, it is heated and poured into huge structures called distillation columns. There, the oil is boiled so that it will **evaporate** (turn into gas). The gas rises up the column. As it gets higher, the gas cools off.

Roads are paved with asphalt, a crude oil product.

Paving playgrounds

Asphalt is a product of crude oil. Asphalt is dark black or brown and very sticky. Because it is waterproof, it is used for paving roads, playgrounds, and car parks.

The crude oil contains many different parts called **fractions**. Some boil at higher temperatures, and others boil at lower temperatures. Fractions that boil at higher temperatures will also **condense** (turn back into a liquid) more easily. These parts condense before they reach the top of the column. Other fractions rise higher in the column before they condense.

When different fractions condense, they are collected in big trays and run out of pipes in the column's side. These pipes are placed at different heights on the column in order to catch the different fractions.

Crude oil is separated into parts called fractions in a distillation column.

crude
oil

cooler

gas (20°C)

naptha (40°C)

petrol (70°C)

kerosene (120°C)

gas oil or diesel (200°C)

lubricating oil (250°C)

heavy gas oil (300°C)

hotter

residue (600°C)

boiler
(super-heated steam)

distillation
column

Distillation of crude oil

Crude oil products

When oil is changed into different chemicals, they are called **petrochemicals**. Many different products are made from these. It may surprise you to know how many petrochemicals you use every day. Plastic is a petrochemical. Every time you put on a pair of sunglasses, brush your teeth, or use a pen, you are using a product that came from **crude oil**.

Crude oil was used to make these plastic sunglasses.

Bright balloons floating at parties and the candles on a birthday cake were produced from oil products. Even water bottles and crayons may come from products that started with oil. From bubble gum to bubble bath liquid, crude oil products are part of our everyday lives.

This woman's jacket is made from polyester, a fabric which is great for keeping people warm outdoors.

Made from oil

There are thousands of petrochemical products. Some of the clothes you wear may be made of polyester, a material that is made from petrochemicals. Your carpet may be made from these as well.

From petrol to energy

One important product that comes from **crude oil** is petrol, a fuel that we use to make our cars run. A fuel is something that makes energy. Petrol is the fuel that is burned to power car engines.

Air mixes with petrol inside the engine. A spark plug sets this mixture on fire. This makes the petrol expand and release energy. It pushes a piston, which turns a rod in the engine. This sets the engine into action. Often four or more pistons take turns doing this work. The burning of the petrol creates the power to make the car move forwards.

When petrol burns inside a car's engine, it releases energy that makes the car run. The waste produced, called exhaust, is then released from the engine.

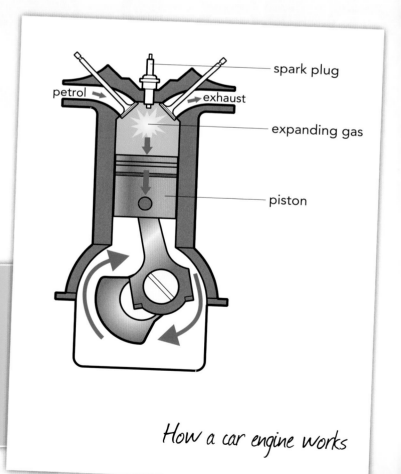

petrol

spark plug

exhaust

expanding gas

piston

How a car engine works

After oil is pumped up from the ground it is often stored in large tanks.

Oil storage tanks in Saudi Arabia

Who produces oil?

The world's greatest producer of oil is Saudi Arabia, a country in the Middle East. Other countries that produce large amounts of oil include Russia, the United States, and Iran.

Running low?

Sources of energy like oil, coal, and **natural gas** are called **fossil fuels**. They originally came from the remains of living things that died. Fossil fuels have been providing energy for people for thousands of years.

However, fossil fuels will not be around forever. They take millions of years to form inside Earth. But we are using them up much more quickly than the time they take to re-form. One day we will run out of these fossil fuels.

Smoke from this factory pollutes the air.

A spill left this penguin covered in oil. Spills can harm or even kill animals.

The dangers of fossil fuels

Some people also worry that burning fossil fuels **pollutes** (dirties) the planet. When these fuels burn, they give off gases. These gases can be bad for our lungs and harmful to the environment.

Sometimes oil transported in ships can accidentally spill into the ocean. These oil spills can destroy animal life. Some people argue that drilling ruins the natural beauty of places such as Alaska.

Future fuels

Scientists know that **fossil fuels** like oil will run out one day. They are looking for other ways to create energy for cars, aeroplanes, and other things we use every day. Scientists are now studying sources of renewable energy. These are types of energy that will not run out like fossil fuels.

Solar power is energy from the Sun that is turned into electricity. Wind power uses the energy of blowing wind. The blowing wind spins big blades on a wind turbine, which produces electricity.

These spinning blades create wind power that will not run out like fossil fuels.

Wind turbines, United Kingdom

This geothermal power station is natural and clean.

Geothermal power station, Iceland

Power from Earth's heat

Another source of energy is geothermal energy, which comes from the heat inside Earth. *Geo* means "earth" and *therm* means "heat". Geothermal energy is used in Iceland, where heat comes from inside Earth through its many volcanoes. It is also used in Mammoth Lakes, California, USA, where there are many hot springs. Geothermal energy is natural, clean, and cheaper than fossil fuels.

Glossary

asphalt crude oil product that is used for paving roads, playgrounds, and car parks

bacteria microscopic living things. Some bacteria can break down the remains of dead plants and animals.

condense turn from a gas into a liquid

crude oil oil when it first comes out of the ground. Crude oil is also called petroleum.

drilling rig machine that drills holes into the ground to find oil

evaporate turn from a liquid into a gas

food chain way that scientists group plants and animals according to what they eat. Plants are eaten by animals. These animals are then eaten by other animals, and so on.

fossil fuel fuel that originally comes from the remains of living things that have died

fraction part of crude oil that has been separated for use in different products

microscopic something so tiny that it can only be seen with a microscope

natural gas gas that forms when bacteria break down dead plankton. Natural gas has many uses, like heating houses and powering clothes dryers, water heaters, and ovens.

oil derrick tall tower with tools inside to drill into the ground for oil

oil well deep hole with machinery that brings oil out of the ground

petrochemical chemical made from crude oil

petroleum oil when it first comes out of the ground. Petroleum is also called crude oil.

plankton microscopic plants and animals that float on top of oceans

pollute make dirty

process change something to make it into a different product

refinery place where crude oil is changed into useful products

sedimentary rock rock that is made up of layers of bits of sand and rocks that fall to the bottom of oceans and lakes. After millions of years, the layers get pressed down and become hard rock.

Find out more

Books

Do you still have questions about how oil is formed and processed? There is much more to learn about this fascinating topic. You can find out more by picking up some of these books from your local library:

Energy from Fossil Fuels (Essential Energy), Robert Snedden (Heinemann Library, 2006)

Oil (Eyewitness Books), John Farndon (Dorling Kindersley, 2007)

The Story Behind Oil (True Stories), Heidi Moore (Heinemann Library, 2009)

Websites

Take a tour of an oil refinery:
http://resources.schoolscience.co.uk/ExxonMobil/vv2/ftpanos.html

Explore this young person's guide to oil and gas:
www.worldpetroleum.org/education/ip1/ip1.html

Take a look at the animation of oil drilling:
www.bbc.co.uk/schools/ks3bitesize/science/chemistry/chem_react_6.shtml

Index